Contents

Introduction

This is a supplementary guide and should be read alongside other guides in the Fire Safety Risk Assessment series.

It provides additional information on accessibility and means of escape for disabled people.

The document can be used to assist in completing the record of significant findings and should include a detailed account of measures that are in place to facilitate and assist disabled people to leave the building.

The appendices provide examples and information to help carry out the assessment and record Personal Emergency Escape Plans (PEEPs).

Technical terms are explained in the glossary.

Where reference is made to British Standards or other standards provided by other bodies the standards referred to are intended for guidance only. Reference to any particular standard is not intended to confer a presumption of conformity with the requirements of the Regulatory Reform (Fire Safety) Order 2005.

1 Background

1.1 Legal overview

The Fire and Rescue Service's role in fire evacuation is that of ensuring that the means of escape in case of fire and associated fire safety measures provided for **all** people who may be in a building are both adequate and reasonable, taking into account the circumstances of each particular case. Under current fire safety legislation it is the responsibility of the person(s) having responsibility for the building to provide a fire safety risk assessment that includes an emergency evacuation plan for all people likely to be in the premises, including disabled people, and how that plan will be implemented.

Such an evacuation plan should not rely upon the intervention of the Fire and Rescue Service to make it work. In the case of multi-occupancy buildings, responsibility may rest with a number of persons for each occupying organisation and with the owners of the building. It is important that they co-operate and co-ordinate evacuation plans with each other. This could present a particular problem in multi-occupancy buildings when the different escape plans and strategies need to be co-ordinated from a central point.

The Disability Discrimination Act 1995 (DDA) does not make any change to these requirements: it underpins the current fire safety legislation in England and Wales – the Regulatory Reform (Fire Safety) Order 2005 – by requiring that employers or organisations providing services to the public take responsibility for ensuring that all people, including disabled people, can leave the building they control safely in the event of a fire.

Where an employer or a service provider does not make provision for the safe evacuation of disabled people from its premises, this may be viewed as discrimination. It may also constitute a failure to comply with the requirements of the fire safety legislation mentioned above.

Public bodies have an additional duty, called the Disability Equality Duty (DED), which from December 2006 requires them to proactively promote the equality of disabled people. This will require them to do even more to ensure that disabled people do not face discrimination by not being provided with a safe evacuation plan from a building.

This document provides guidance on how organisations can ensure the safe evacuation of disabled people from their premises.

1.2 Management practice

The DDA requires organisations to review their policies, practices and procedures in order to ensure that they do not discriminate against disabled people, and to take steps to overcome any physical barriers that make it impossible or unreasonably difficult for a disabled person to use a service. Operational procedures, for example those that require all visitors to park away from a building, have had to be amended to allow disabled people to park close to the main entrance.

Equally, the practice of locking the side swing door adjacent to a revolving door is likely to be unlawful under this part of the Act. Such an act may also constitute an offence under current fire safety legislation. These are examples of how the DDA changed how companies manage public access.

However, attention was focused on getting into premises, when, of course, if one is going to enable disabled people to fully use the building, one also needs to enable them to leave safely. The safe evacuation of disabled people is a problematic area for policy makers and one that has not received sufficient attention to date.

It is important that both building managers and disabled people understand that planning for means of escape is about planning for exceptional circumstances (i.e. not an everyday event). When writing escape plans that include disabled people, there is sometimes a tendency to overplay the safety issue to the detriment of the independence and dignity of disabled people. The purpose of this guidance is to provide you with clear information so that your organisation is able to deal with these issues in a practical, equality-based manner.

It should also be remembered that what a disabled person is prepared to do in exceptional circumstances may differ significantly from what they can reasonably manage in their everyday activities. Escape plans for disabled people should be prepared with the view that what is required is for 'the real thing'. The level of effort required of a disabled person may not be acceptable for a practice or false alarm or in everyday activities. The procedures put in place should take account of this and allow for simulation in the case of fire drills or other emergency evacuation practices.

Good housekeeping standards and management procedures will reduce the incidence of false alarms.

1.3 Reducing unnecessary escapes

Some disabled people are put at a great risk when carry-down procedures of any kind are used. It is therefore necessary for the evacuation policy to include a method of reducing or removing the need to escape for a false alarm. It is likely that many more disabled people will be willing to facilitate their own escape when they know that this is not going to be required of them during a practice or for a false alarm.

Good communication with disabled people about the fire or emergency evacuation process is vital to ensure its success and to reduce the need for emergency escapes except in exceptional circumstances.

1.4 Personal Emergency Evacuation Plans (PEEPs) for employees and regular visitors

Where staff and regular visitors to a building require a plan, they can be provided with an individual plan through the human resources department or building manager. The plan must be tailored to their individual needs and is likely to give detailed information on their movements during an escape. It is also possible that there will be some building adaptation to facilitate their escape and to reduce the need for personal assistance.

Example

A health club has a regular member who finds the stairs difficult. During their induction, the fitness instructor discusses their escape needs. An evacuation chair is provided at gym level. All instructors are trained in the use of the chair and they are introduced to the member.

1.5 Standard plans for occasional visitors

This guidance provides advice on a wide range of options for ensuring the safe evacuation of disabled people. These options contain some standard elements, but these can of course be adapted to suit particular organisations. In order to provide suitable means of escape for visitors, a set of standard escape options should be adopted by the organisation.

A standard plan is used where there are visitors or casual users of the building who may be present infrequently or on only one occasion. The provision of standard PEEPs takes account of the following:

• the disabled person's movements within the building;

• the operational procedures within the building;

• the types of escape that can be made available;

• the building systems, e.g. the fire alarm; and

• the existing egress plan.

Standard evacuation plans are written procedures that can be used as options for disabled people to choose from. They are held at the reception points within the building and are advertised and offered to people as part of the entry/reception procedures.

This is an extension of the process of signing into a building and being given a visitor badge with the escape procedures on the back of it. A disabled person requiring assisted escape is offered options for their assistance and is given suitable instructions.

It is understood by most people that when a fire alarm is activated they must all leave the building by the nearest exit, as quickly as possible, and reach a place of ultimate safety. The management of the building is required to keep escape routes clear and free from obstruction and to ensure that exits are readily available for use on quick-release devices which also offer protection from unwanted or illegal entry. However, everyone using a building for whatever purpose should also take some responsibility for their own safety wherever possible.

This responsibility also applies to disabled people, therefore disabled people can be expected to identify themselves when they are informed of the availability of a choice of evacuation plan and co-operate by giving any information necessary for the safe execution of the plan.

Example

A visitor approaches reception, where there is a clear sign indicating the provision of a PEEP system. The visitor has a visual impairment and therefore requires information about the escape routes. The building operates a policy of the meeting organiser being responsible for visitors if an escape is necessary. The receptionist explains the process for obtaining support.

The visually impaired person makes the meeting organiser aware of the need for assistance. All staff are trained in disability escape etiquette. Prior to the start of the meeting he/she points out the escape routes and offers to assist if necessary.

1.6   Unknown or uncontrolled visitors

Where there are people within the building who do not pass a reception point or are not controlled, such as in a shopping centre, library or theatre, it is more difficult to gather information prior to the need to escape. In these instances a system of standard PEEPs should also be implemented and advertised.

Training for staff is vital in this case as they will have to provide assistance and advice to disabled users of the building as the incident develops. The plans to enable them to leave safely in the event of an incident will require pre-planning. Staff will need to understand all the options within the matrix (contained in Appendix 1) and be able to communicate these effectively to disabled people at the time of escape. In order to do this, they should receive disability escape etiquette training.

In large, multi-occupancy buildings, it will be essential for each organisation to ensure that suitable training is provided to all their staff. Such a training requirement should form part of their fire safety risk assessment.

Example

A museum is required to evacuate due to an alert in one of the galleries. There are a number of wheelchair users present. The museum has a high standard of compartmentation due to the need to protect the exhibits. This is an advantage in an escape situation and staff members have been trained to understand the safety implications of this fire safety feature. The communications process set up as part of the escape procedures for staff tells them where the alarm has been raised. They can then direct people who cannot use stairs away from the alarm point to a safer part of the building.

1.7   Small buildings

In larger buildings, the building systems and options are likely to provide more options than in smaller buildings. However, in smaller buildings there will be fewer people and greater opportunity to communicate. A standard set of plans should be developed in the same way as for a larger building.

2 Communication

2.1   Consultation

When producing an evacuation plan which includes disabled people, it should be remembered that normally people cannot be expected to react exactly as planned in any emergency. It is generally accepted that, unless guided by trained staff, most non-disabled people (including those who may have worked for years in a building) will make their way to the exit that they are familiar with, rather than to the most suitable escape route. Provision of a fully integrated PEEP system will benefit all groups of people and will identify any weaknesses in existing evacuation plans. Therefore, it should not be considered a burden on the evacuation plan, but an opportunity to improve safety for all people using the building.

The different groups of people who should be considered and are likely to be present in a building are as follows:

• staff;

• contractors;

• visitors;

• residents;

• students; and

• customers – individuals and groups (hiring out of rooms, public events, etc.).

Each of these groups has a different role to play and it is likely that the methods of contacting them will need to be different. This will require a communications strategy that involves the people responsible for managing the use of the building. It will also require those people to work together to ensure that a joined-up and co-ordinated approach is taken.

The method of making contact with disabled people and the type of evacuation plan they are provided with will differ depending on the function that they are fulfilling within the building. The type of building will also influence the type of plan.

The general population will follow the escape routes or make their way out by the way they came in, but disabled people who require their escape to be facilitated will need to be considered in more depth in the general plan. Disabled people will need to have more information about the options available to them. In some instances, they will need to be allocated people to assist their escape; however, the aim should be to facilitate disabled people's independent escape as far as possible.

2.2   Making contact and defining roles

It is easier to contact staff and regular visitors to the building and, generally, this will be done through the personnel procedures and general management systems. It will also be easier to prepare detailed escape plans for these disabled people. It is also likely that volunteers to provide assistance to disabled people can be easily recruited from their peer group.

Where standard PEEPs are used and disabled visitors are not available to consult with in person when setting up the system, it is appropriate to consult local disabled people's organisations.

It will be necessary to allocate responsibility for the provision of a suitable plan for each group of people to a relevant member of the staff team. A list of building users and appropriate staff who will need to be involved is provided below.

2.2.1   Staff

The responsible person will be responsible for ensuring that staff are provided with suitable escape plans. In creating suitable escape plans, the responsible person would be advised to involve human resources departments, where they exist, or line managers, who may hold information relating to disabled employees and may also have responsibility for training and the development of staff skills.

Staff have a vital role in communicating the evacuation plan to disabled visitors, and to fulfil this role effectively they will be required to undergo **disability escape etiquette** training. This consultation and planning process should be introduced on induction and be reviewed regularly as appropriate. Information should also be provided within the staff handbook. A system is required to ensure that plans are regularly updated (see Appendix 4).

2.2.2   Contractors

Where there are contractors working in the building, the responsible person has overall responsibility for their safety in case of fire; however, this may often be delegated to a competent person in the department they are working for. The competent person should ensure that steps are taken where necessary to ensure that they are provided with a suitable escape plan chosen from the standard set of plans for the building.

2.2.3   Residents

Where sleeping accommodation is provided, e.g. in a hotel, part of the booking-in procedure should include the offer of a suitable escape plan. Additional accessible information is required in each room, adjacent to the evacuation procedures for all residents.

In hostel accommodation or student dwellings, etc. suitable PEEPs should be provided by the accommodation manager based on the standard set of plans for the building.

2.2.4   Students/pupils

When a child or student is enrolled, their escape plan should be developed as part of the admissions process. Care should be taken that all disabled children or students are provided with a plan if they need one, even if they are provided with a statement or not.

2.2.5   Visitors – individual

Individual visitors to a building may fall into two groups: those who are invited to a building, such as sales representatives; and casual visitors who attend of their own volition, such as clients attending to discuss issues with members of staff.

A system of standard plans should be created. For invited visitors, the plans could be put in place prior to the meeting, or they could be presented to casual visitors when they book in at reception.

2.2.6   Visitors – groups

Part of the booking procedures for groups should include provision of standard plans. Where there are a large number of disabled people, it may be acceptable for the party organiser to play a role in the provision of suitable escape plans. Booking administration should facilitate this.

2.2.7   Casual visitors

In public access buildings, it may be impossible to know how many disabled people are present at any one time or their level of disability. In such cases, responsibility for evacuating them safely in the event of an emergency will rest with staff and building managers. It is important, therefore, that staff and managers fully understand the evacuation plan and fire safety strategy for the building so that they can render maximum assistance to disabled people, irrespective of the nature of their impairment. Staff and management training and empowerment are crucial factors in this planning process.

Example
A serious fire occurred in a nightclub in a major city centre in the UK. Due to the prompt and effective action of staff and managers in evacuating customers from the building, 500 people were successfully evacuated safely into surrounding streets.

2.3   The communications process

A communications process is required so that there is suitable support for the evacuation plan system at each level of the building. It is necessary to consider the following steps within a plan.

2.3.1   Co-ordination

A co-ordinating role is necessary in order to ensure that any plans provided are understood throughout the organisation. Overall responsibility for this role rests with the responsible person(s); however, in practice this is likely to be delegated to a competent person from the human resources department or safety services. Different members of the organisation will be appointed as competent persons and will be responsible for ensuring that there is provision for means of escape for disabled people using the service that they provide. The competent persons will report back to the co-ordinator.

2.3.2   Technical building information

Technical information is also required about the building systems, the fire safety systems and the fire safety strategy for each building. This information should be made available to all of the people who are to be part of the escape plan. For instance, if the building has suitable fire compartmentation to allow horizontal evacuation into another fire compartment, people operating the plan should understand why this is possible.

2.3.3   Staff provision

a) Human resources departments will normally have the day-to-day responsibility for staff and should ensure that all staff are offered a suitable escape plan during their induction process or where there is any change to the person's ability to make their way out of the building.

b) The head of each department will normally be responsible for their own staff and should arrange the provision of a PEEP for each person requiring one. It may be necessary to provide a plan for each building and room that they visit.

c) A disability contact, if there is one, and if not the line manager or competent person in each department, should take on this role and ensure that the PEEPs for the staff under their care are kept up to date by contacting/reminding the department.

2.3.4   Visitors to the building

An appropriate contact point for each group of people visiting a building should be established. For instance, this may be:

• the main reception point; or

• via the meeting booking procedure; or

• via the person or department that they are visiting.

This will depend of the nature of the organisation.

2.3.5 Additional support from security and portering staff

Where there are security and portering services, these can provide a support role and allocate standard plans for visitors. They may also provide assistance in some instances. It is important that these members of staff are provided with suitable training and fully understand their role, particularly where their function is outsourced.

2.3.6 Training and recruitment of volunteers

In some instances it may be necessary to recruit and train additional staff to provide assistance during an escape. In considering staff who may provide assistance in an evacuation, it is important to take account of their work-time availability, location in a building or on a site, and whether they are employees of another company providing an outsourced facility. Another consideration in utilising outsourced employees is the need to ensure that their managers are fully in agreement with their involvement in an emergency plan and that the person concerned is fully conversant with the work culture and policies of the workplace or site.

2.3.7 Functions and conferences

Function/conference organisers will be responsible for ensuring that disabled people attending conferences or meetings within the building are provided with a suitable plan. It is important that conference fliers and booking forms inform delegates about the building systems.

2.3.8 Meetings

When a room is booked, a standard procedure should be to check if there are disabled people attending. If so, a suitable escape plan will be required.

2.3.9 Residents

When a disabled person is allocated a room (whether it is specially adapted for them or not), a suitable escape plan should also be provided. Some disabled people who use hotel or other residential sleeping accommodation may not need an adapted room but may need support to escape, e.g. blind people. Therefore, a clear sign is required to be displayed at reception and alongside the escape instructions in each room.

Escape instructions displayed in each room should be made available in other accessible formats, for example the receptionist could explain the instructions after the person has checked in.

Standard plans for the building should be allocated to visitors by the reception service in that building.

2.3.10 Training programmes

In order to ensure that the system runs smoothly, it is important to introduce a regular training plan. The following is an example diary, inlcuding training dates.

| Year 1 | | | | | | | | | | | | Year 2 |
Jan	Feb	Mar	Apr	May	Jun	Jul	Aug	Sep	Oct	Nov	Dec	Jan
MOE training	Carry-down training	Mock-up				MOE training	Carry-down training	Mock-up				MOE training

MOE – Means of escape
Mock-up – Simulation of disabled people's escape procedures
Carry-down – All types of escape that include evacuation chairs, manual handling training, disability evacuation etiquette training

N.B. Staff involved in the escape plan should feel confident in their skills and disabled people should feel that they can trust the process.

2.3.11 Budgets

It may be appropriate to allocate a budget to improve the emergency escape provision within the building.

3 The process

3.1 Interviewing staff

Once the person responsible for their plan has contacted the disabled person, an interview should be organised to establish suitable evacuation procedures.

A suitable plan should be negotiated, taking into consideration what the building, management and disabled person can offer. It should not be automatically assumed that a disabled person cannot leave the building independently. It is recommended that disabled people are consulted about their evacuation plan. They should be given information about the building systems and their opinions and experience should be both sought and respected.

The appropriate time required to make the disabled person's escape should be identified. Disabled people should not automatically be required to wait for the main flow of escape to be completed. However, if they are likely to cause obstruction for other people leaving the building, it will be safer for everyone if they follow the main flow of people.

Wherever possible, the escape plan should accommodate both fast and slow-moving people. However, where the person may need to rest or they feel threatened by people behind them, it may be appropriate to design a plan that allows for this, e.g. resting in refuges provided along the route.

The matrix in Appendix 1 gives the options that are suitable for most disabled people. This should be used alongside the information that is provided about the building. The two can be matched together to form either a standard evacuation plan or an individual plan.

Example

A visually impaired person is working in a building that has a main entrance incorporating the main stairway and one additional escape stair at the other end of the building. The escape stair has suitable handrails and step edge markings. The person is familiar with the building and has been shown where the escape stair is. They elect to make their own way out of the building because the access provision in the escape stair is adequate.

Most disabled people are likely to have a very clear idea of what it will take to get out of the building. In some instances, the person will be able to facilitate their own escape if suitable aids and adaptations have been provided. The responsible or competent person working with the disabled person to write the plan should not make assumptions about the abilities of the disabled person. They are likely to know what they can achieve.

Where a person can make their escape unaided, it may take them longer than the three minutes generally accepted as the time taken for non-disabled people to make their escape in case of fire. They should be given the opportunity to take the safest route, which offers them the longest period of safety, for instance through to an adjacent fire compartment, which has a one-hour fire-resisting rating, and then down the escape stair, which has a 30-minute fire-resisting rating.

Where staff assistance is required, sometimes this will be by staff within the department concerned. Where local staff are not available, contact should be made with the responsible person for the building so that a suitable alternative option can be set up.

3.2 Contacting unknown visitors

It is much more difficult to organise an escape plan for people who are casually visiting the building or for people who are using the service on a one-off basis. However, by assessing the types of escape that can be provided within the building in the same way as for a known population, it is easier to address their needs.

Once the escape options are known, staff should be trained to implement them at the time of an escape. This will require organisation and practice. Using fire drills that involve disabled members of the public is not advised as it may put the disabled people at risk from injury unnecessarily. Regular simulated practice should take place alongside moving, handling and disability evacuation etiquette training.

3.3 Recruitment and training

Sometimes it can be difficult to recruit volunteers as they will want to be sure that their own safety is not compromised by helping the disabled person to escape. It may be necessary to raise the awareness of staff prior to the recruitment of volunteers so that they understand that their own safety will not be compromised.

Clear information should be provided to volunteers about facilitated and assisted escape systems. It would also be supportive to potential volunteers to assure them of the organisation's commitment to their continued training and support. In some instances it will be necessary to provide a session for potential volunteers so that they feel more comfortable about coming forward. Accreditation and possible remuneration for volunteering for this training may also be introduced, in the same way as exists for some first aid staff.

The training provided should include disability awareness, disability evacuation etiquette, and moving, lifting and handling techniques.

3.4   Practice

Practice for PEEPs will depend on the type of escape required. Generally, escape plans should be practised on a regular basis and at least every six months. However, some systems will need testing more frequently than that, for instance paging systems.

All the people involved in the escape plan should take part; however, it may be more appropriate to simulate carry-down so as not to cause unnecessary risk to the disabled person.

Where a disabled person has elected to make an exceptional effort to get out unaided, it is not practical for them to practise; however, timing a short section of the escape will help in establishing how long a full escape might take.

People with a learning difficulty may need to practise their routes for escape on a monthly basis. If so, this should be written into their PEEP.

3.5   Co-ordinated information

Once each plan is written, it should be passed on to the responsible person(s) within the building. This will ensure that the plans for each premises and its occupants in a building can be co-ordinated. This is especially important where there is potentially a high number of people to be evacuated to ensure that there is no conflict.

Under fire safety legislation, the responsible person has overall responsibility for ensuring that all emergency plans are updated as necessary and whenever the fire risk in the building changes.

Where this responsibility is delegated to a competent person, that person should ensure that it is not overlooked. It is important that, should that person leave or be away on long-term sick leave or maternity leave, their role is allocated to another suitably trained person either permanently or for the period of their absence. Disabled people should be advised to tell their nominated person of any change in their circumstances.

4 People's preferred options for escape

4.1 Negotiate 'reasonable adjustments'

Generally, disabled people are no different from anyone else in that they prefer to be in control of their own escape. The DDA requires that adaptations may be made to physical features of buildings to enable them to be used more easily by disabled people. However, the DDA recognises that it may not be possible to provide full access. The minimum requirement is difficult to outline, but a good guide would be to use the specifications set out in BS 8300. These can be considered a measure of accessibility under the DDA and can be considered desirable features for means of escape.

Sometimes there may be difficulties when managers are trying to introduce PEEP systems. Disabled staff or visitors can sometimes expect the provision of items such as lifts where it is not feasible to provide these. It is important that where such conflict arises both parties take a realistic view of the situation. Managers should be prepared to discuss with disabled people what options there are and what provision they can make. Disabled people also need to understand the limits of reasonableness set out by the DDA.

The following statements should be considered as part of the negotiation procedure:

- Health and safety legislation requires building managers to ensure the safety of staff and visitors to a workplace.

- The Regulatory Reform (Fire Safety) Order 2005 requires that all people using the building be provided with adequate means of escape in case of fire. This includes a suitable escape plan.

- There is also a responsibility for all staff using the building to be aware of and to practise the escape procedures periodically. It works on the principle that people are responsible for their own escape, which will be facilitated by the building management and provided for by the responsible person.

This implies that disabled people also have a responsibility to co-operate with the provisos of their own escape plans and to facilitate their escape. Often there is reluctance on the part of disabled people to volunteer information about what they could achieve in a one-off escape situation. In order for disabled people to be willing to volunteer this information, responsible persons and building managers should take the right approach, recognising disabled people's dignity and right to independent access and evacuation, and they should provide as much information as possible to everyone about the plans for disabled people. This will encourage disabled people to be more frank in their approach to establishing their own escape plan.

Some negotiation skills, sensitivity and level of discernment are required here on the part of the person carrying out a PEEP. (Disabled people may feel pressured to do more physically than they would generally be able to achieve, or they may be afraid that back-up systems and support will not be made available to them.) Training is essential.

It should be made clear to disabled people (while working with them to develop a suitable plan) that the circumstances of escape are considered to be exceptional. That means solutions that may not be appropriate in most circumstances could be used, such as allowing a disabled person to move down the stairs on their bottom. It would not be acceptable for them to do this in any other circumstances. The disabled person may need assurances that, if they volunteer what they might do in an emergency, this will not constitute grounds for the removal of any support at other times.

Not all people who have an apparent impairment will require an assisted escape plan. Also, it should not be assumed that people with invisible impairments and who normally would not have an access problem will not require assistance in an emergency situation. This may be caused by the fact that current guidance on means of escape in case of fire is not necessarily consistent with access standards, e.g. lift access to upper floors without an evacuation lift provision, edge marking of stairs.

All staff should be given the opportunity to have a PEEP at induction. The reason for this is that some people may have difficulty in evacuation situations that they would not have normally, e.g. people who have asthma may be affected in smoky conditions caused by a fire, or people might be affected by the stress of an emergency situation.

4.2 Mobility impaired people

There is a vast range of people who fit into this category. Issues relating to this group of people may also be relevant for people who have heart disease, asthma or heart conditions.

The preferred options for escape of people with mobility impairments are by horizontal evacuation to outside the building, horizontal evacuation into another fire compartment, or fire evacuation lift, eventually arriving at a place of ultimate safety outside the building. This is the preferable option for disabled people. Within this group, many people will be able to manage stairs and to walk longer distances, especially if short rest periods are built into the escape procedure.

A possible facilitating measure may be the provision of suitable handrails. Information regarding the position of the fire is also useful so that there are no false starts or the necessity to change direction during the escape.

It should also be remembered that escape from the building within two to three minutes may not be possible for this group of people. It may be advisable to explain which escape routes have a degree of fire and smoke resistance and how the building is compartmented.

The level of fire protection available and identification of elements such as compartmentation and fire alarm zoning within the building will help buy the time required for disabled people to either facilitate their own escape or leave with assistance.

4.3 Wheelchair users

This group of people is considered most at risk in terms of escape. However, in some instances, a person who frequently uses a wheelchair may be able to walk slightly and therefore be able to assist with their own escape or even facilitate independent escape. It is essential that the disabled person is asked the relevant questions tactfully and in a way that produces the best escape plan.

Assumptions should not be made about the abilities of wheelchair users and they should not be excluded from a building because of false assumptions about their ability to leave the building safely.

The preferred method of escape by most wheelchair users is horizontally to another fire compartment, or to outside the building, or vertically by the use of an evacuation or fire-fighting lift. If these options are not available, or not in operation, it may be necessary to carry a person up or down an escape stair. Carry-down can be achieved in a number of ways, as set out below.

4.4 Carry-down procedures

4.4.1 Evacuation chairs

This looks like a deckchair with skis and wheels underneath. When placed on the stairway it slides down the stair. There are wheels at the back that facilitate movement on the flat, but they are not suitable for long distances.

An evacuation chair is operated by one or two people and requires training and practice to use. Disabled people may not feel confident using these chairs and it is not always possible for wheelchair users to transfer into an evacuation chair or to maintain a sitting position once seated in one. Therefore, evacuation chairs should not be considered as an automatic solution to the escape requirements of wheelchair users.

It is unlikely that an evacuation chair will be of much use unless both the user and the operator are well trained and familiar with the piece of equipment. It is essential that when they are purchased a suitable training system is also implemented. Regular practices should also take place. In most instances, these may not need to include the disabled person, although some may wish to practise being moved in the evacuation chair. It is more appropriate for the people who are trained to operate the evacuation chair to take it in turns during practices rather than involve the disabled person. This will also increase their confidence in using the equipment. Using an evacuation chair may put the disabled person at risk from injury, so it is best to limit their use by disabled people to the real thing.

4.4.2   Carry-down in the person's own wheelchair

It is possible to move a person down a stairway in a number of ways using their own chair as an aid.

Carry-down by two, three or four people can be done by holding the wheelchair at one of the fixed points situated in each corner of the wheelchair. The team then lifts the wheelchair and moves up or down the stairway. Many wheelchair users will be able to point this out.

4.4.3   Carry-down using an office chair

This can be used when a person does not have a wheelchair that is suitable for carry-down, for example a large motorised chair.

Any stable office chair can be used, although preferably it would be one with armrests. The carry-down is facilitated in the same way as when using a wheelchair.

4.4.4   Carry-down using 'wheelies'

With some wheelchairs it is possible to tilt the chair on its axis so that it is virtually weightless on the stair. With either one or two people holding onto the chair by a fixed point at the rear, the wheelchair can be manoeuvred down the stairs, allowing the weight of the person to carry the chair down the stairs. Some wheelchair users are able to make this manoeuvre unaided; however, these people are in a minority, and, in any case, the manoeuvre is really only practical on a short flight of stairs.

None of the above techniques should be attempted without appropriate training. All types of carry-down escape techniques require a risk assessment and professional moving and handling training for the operators.

When designing the escape plan, remember to consider what is practical and achievable in exceptional circumstances rather than what might be achieved in normal day-to-day activity.

4.4.5   The interview

When writing a plan with someone who has a mobility impairment, or who uses either an electrically or manually powered wheelchair, the following information should be obtained:

• which routes have handrails provided;

- how far the distance of travel is on particular routes;

- the degree of fire compartmentalisation within the building and the exact location of the fire compartments;

- the provision of evacuation chairs;

- which staircases are provided with handrails and what side of the stair they are situated on;

- the opportunity to use lifts and lift locations; and

- what staff assistance may be available.

Questions to ask during the interview include:

- Can you walk aided/unaided down the stairs?

- How far can you walk unaided?

- Can you slide down the stairs?

- How many flights can you manage?

- Would this be increased if assistance were made available?

- How many people would you need to assist you?

- How many times might they need to stop to rest?

- Would handrails be of use in assisting your escape?

- Are there positions along the escape route where handrails or other aids might assist you?

- How might your mobility be worsened, e.g. by smoke, etc.?

- Is your wheelchair electric or manual?

Once this process has taken place, some people will decide that they can facilitate their own escape using the systems within the building. Others will decide that they require assistance from one or more people.

4.5 Electrically powered wheelchairs

People with limited mobility – possibly heavy

People who use electrically powered wheelchairs may have less mobility than people who use manual chairs. However, there may be exceptions to this rule, so it is important to consult the disabled person wherever possible.

This group of people is likely to require much more assistance when leaving the building. It is wise for the responsible person or building manager to facilitate the independent escape of all other groups of disabled people in order to ensure that there is sufficient staff to assist this group.

It is impractical to expect that this group of people will be able to take their chair with them, due to its weight and size. They will need to leave their chair in the building if there is no suitable lift to facilitate their escape. This will mean that some other method of carrying them down the stairs will be required. This may be a piece of equipment such as an evacuation chair.

There are other types of mechanical equipment that exist to move people up or down stairs; however, timing and obstructing the escape of others are prime considerations if thinking about using this type of equipment for evacuation purposes.

An important issue to consider when planning means of escape for people who require carry-down by four people is that the width of the stair will need to be sufficient for all of the team to move freely and safely.

4.6 Hearing impaired and deaf people

Hearing impaired and deaf people need to know that there is an escape in progress. Where only an audible fire alarm system is present, they may not be able to hear the alarm or any information being broadcast by PA systems. However, if sound enhancement systems are provided within the building, it may be possible to transmit the message through that system, e.g. via a hearing loop or radio paging receiver.

The preferred options to alert hearing impaired people that an emergency exists and an evacuation is about to occur are the use of flashing beacons installed as part of the fire alarm system and the use of a paging system. However, these cannot always be provided. Where this is not possible, there is a range of other auxiliary aids to provide this information.

4.6.1 Information required

When writing a plan with someone who has a hearing impairment or who is deaf, information should be obtained on whether any of the following pieces of equipment are available:

• visual alarm system;

• MSN text messaging;

• office intranet;

• telephone network – textphone;

• vibrating pager;

• team member;

• fire wardens;

• appointed buddy; or

• local beacon.

All pagers and other equipment should be tested regularly to ensure that they work.

4.6.2   Staff training

Where other staff are used to alert hearing impaired or deaf people that they need to leave the building, they should be trained in deaf awareness. Often floor wardens sweep the building to ensure that there is no one left on the floor. These staff can be trained to look for signs that a hearing impaired person is present who may not have heard the alarm.

A typical situation where this may occur is in single offices, libraries, toilet accommodation or changing rooms. Fire wardens should not expect a vocal call to be sufficient and should be trained to physically check all areas for which they have responsibility, provided it is safe for them to do so.

Staff should also be aware that when a person does not react in a logical manner during the escape procedure they may not have heard the alarm. Shouting louder is unlikely to be the answer. It may be necessary to walk right up to the person and explain what is happening with signs or even a written note or pre-prepared short written instruction.

4.6.3   Fire instructions

It should also be recognised that many hearing impaired and deaf people do not have English as a first language. It is important that a Plain English translation of the fire protocol is provided. It may also be an advantage to this group of people for pictograms to be provided to support the written information. Deaf people may prefer to have instructions explained to them through a British Sign Language (BSL) interpreter.

There are additional issues to consider when writing a plan for a hearing impaired or deaf person.

4.6.4   The interview

The following information should be given to a hearing impaired or deaf person when writing their plan:

• the systems that are available to advise them of an evacuation, e.g. alarm beacon, pager, personal contact, etc.; and

• the technical operation of fire alarms – how to raise the alarm, how to contact the control room, etc.

Ensure that they are aware of the evacuation procedures – where to go, alternative routes, and where to report to after the evacuation.

• The following questions should be asked when writing the plan:

• Do you work alone in the building?

- Do you work out of hours?

- Can you hear the alarm?

- Do you work as part of a team or in a group office environment?

- Do you have a dedicated text number?

- Do you have an email address?

- Are you likely to move around the building?

4.6.5   Lone working

Care should be taken to ensure that hearing impaired or deaf people who are working alone in a building know what is happening. In these instances, it may be imperative that a visual alarm system or vibrating pager system is installed.

Similarly, this is also important where a person is working out of hours and where there may be no other hearing people available to advise them that there is an emergency evacuation in progress. Remember that the evacuation system may be used for purposes other than a fire emergency.

The working hours or working flexibility of hearing impaired or deaf members of staff should not be restricted because inadequate provision for safe evacuation has been made. Such restrictions, if made without full consideration of reasonable adjustments, may amount to unlawful discrimination.

Example
A senior manager who is deaf is required to work late and be in the office early on occasion. The office has a 24-hour security presence and it is necessary for all staff to sign in and out at the security point. An arrangement is made that, should an alarm be raised out of hours when the manager is present, the security guard will contact the manager on their textphone to alert them of the emergency. This was built into the instruction manual on means of escape procedures for all security staff.

4.7   Visually impaired and blind people

People who are visually impaired are helped to escape by the provision of good signage and other orientation clues. It should be noted that most visually impaired people have some sight and that they will be able to use this during the escape in order to make their own way out of the building as part of a crowd. Where the physical circumstances are appropriate, they will have no problems leaving the building.

Some organisations will not have made provision to provide specialist orientation information, for example tactile information and audio signals. Use can be made of existing elements within the building that might help visually impaired people to facilitate their own evacuation. These may be elements of building design, such as good colour contrasts, handrails on escape stairs, step edge markings on escape stairs, colour contrasted or different textured floor coverings on escape routes or way finding information. Where orientation clues are provided, these will further reduce the need for assistance.

However, there will still be a need to inform visually impaired people of the presence of these via the PEEP. Where there is a lack of orientation information, staff assistance will be necessary to provide guidance out of the building.

4.7.1 Orientation information

Improving circulation and orientation can be of great benefit. Logical routes to escape stairs will not only assist visually impaired people but will be of benefit to all users of the building.

Good colour definition and accessible signage will help visually impaired people to use the building. Extending these systems to include the escape routes can reduce the need for assisted escape.

A visually impaired person might not easily locate the exit signs and may not be aware of the travel direction to get out of the building, but they may remember their way out along the route by which they entered the building. Using the escape routes as part of the general circulation space within the building will mean that visually impaired people will become more familiar with these routes and will therefore have more options for making their escape.

4.7.2 Fire instructions

Visually impaired people are not generally able to read the fire escape instructions provided in most buildings, as these are often in very small typefaces. Suitable instructions should be made available in Braille, large print or on audio-tape. It can be useful to provide a tactile map of the escape routes and to provide orientation training to visually impaired staff working in the building, so that they are more aware of the options for escape.

4.7.3 Staff responsibility

Visitors to the building are unlikely to spend time alone. Rather than provide a focused escape plan for each individual person, a philosophy should be adopted that gives staff the responsibility of ensuring that their visitors leave the building safely, whether or not they have a disability. This would be preferable to providing extensive and possibly unappreciated escape training for the casual visitor.

4.7.4 Keeping routes safe

Some other simple measures can be adapted to facilitate visually impaired people in making their escape. They may have difficulty in stairways where there are open risers and these should be avoided on escape routes. Where these are present then there may be a need for assistance or adaptations to the stairs to make them safer. Alternatively, a different stairway may be available.

When office furniture is rearranged and escape routes are affected, it is important that these changes are documented and made known to visually impaired people in the building.

4.7.5   The interview

When writing a plan with someone who has a visual impairment, the following information should be obtained:

• What type of alarm system is available?

• Are the escape routes clearly marked?

• Is there sufficient orientation information?

• Are fire instructions provided in accessible formats?

• Are there step edge markings on the escape stairs?

• Are there handrails on the escape stairs?

• Are risers closed?

• Are there external, open, steel escape routes?

• The questions that should be asked are:

• Do you work alone in the building?

• Do you work out of hours?

• Can you hear the alarm?

• Are you aware of the positions of all the escape routes?

• Can you follow them unaided?

• Do you work as part of a team or in a group office environment?

• Are you likely to move around the building?

Can you read the escape instructions? If not, what format do you need them in?

4.8   People with cognitive disabilities

People with cognitive impairments often have problems comprehending what is happening in an evacuation or may not have the same perception of risk as non-disabled people.

Some people with cognitive impairments may fall into the group having unknown requirements, such as dyslexia, dispraxia and autism. These people may not be aware of their impairement. Many people with learning disability also have other disabilities, some may have mobility difficulty and some may have impaired vision and hearing loss. Sometimes people with cognitive disabilities will move more slowly than the main flow and there may be a need for a slow and fast lane in the escape stair if the stair width allows this.

4.8.1   Orientation information

Orientation information and colour coding of escape routes can also provide a useful tool. Practice of the route options can dramatically reduce the requirement for staff assistance. Practice is essential for this group of people, especially in situations where one person is responsible for a number of others, for example in a classroom situation. Use of escape routes for general circulation is an advantage.

4.8.2   Fire instructions

This group of people may need to have the escape plan read and explained to them. A video or DVD explaining and demonstrating what to do in an emergency can also be an advantage. A photographic explanation of the route can also be useful.

Rather than merely asking what this group of people needs, it may be more relevant to ask what they understand and to develop the plan based on how they will find the escape routes.

4.8.3   Other factors

Sometimes people with cognitive disabilities will move more slowly than the main flow and there may be a need for a slow and fast lane in the escape stair.

It is important to understand that not every person with a cognitive impairment will have a carer or helper with them, so efforts should always be made to enable the disabled person to understand how to leave the building rather than assuming that a carer or helper will undertake this role.

It may not be possible to tell that a person has an impairment that affects their ability to orientate themselves around the building, and staff should be made aware of such possible situations and be tactful when assisting a person who may seem lost or unsure of what to do during an escape.

4.8.4   The interview

When writing a plan with someone who has a cognitive impairment, the following information should be obtained:

• What type of alarm system is available?

• Are the escape routes clearly marked?

• Is there sufficient orientation information?

- Are fire instructions provided in accessible formats?

- Are there step edge markings on the escape stairs?

- Are there handrails on the escape stairs?

- Is it likely that there will be a need for two-speed traffic on the stair? If so, is it wide enough to allow this?

- Are risers closed?

- Are there external, open, steel escape routes?

The questions that should be asked are:

- Do you work alone in the building?

- Do you work out of hours?

- Do you know what the alarm sounds like?

- When you hear the alarm, do you know where to go?

- Do you work as part of a team or in a group office environment?

- Are you likely to move around the building?

- Can you read the escape instructions? Do you understand them? If not, what format do you need them in?

4.9   Unknown requirements

It should not be assumed that because a person has a disability they will need or ask for a PEEP. Some will be confident that they can get out of the building unaided. Conversely, there should also be an opportunity for other people who may not be considered as having a disability to request an escape plan. All staff in a building should be given the opportunity to have a confidential discussion about their escape requirements and be clear that, if they need help, it will be provided. The service provider should adopt an approach that enables people to ask for a plan, when needed, without them feeling that it will affect the provision of that service to them in any other way.

One group of people who may find themselves in the category of 'unknown requirements' is people with epilepsy; however, they are not the only people who may have such an unknown requirement. Many will be able to leave the building unaided in an emergency, but some managers may not understand this.

For example, they may assume that a person with epilepsy will have a seizure due to the fire alarm operating and may collapse in an area where they are on their own (e.g. a toilet cubicle or storeroom) so that no one knows where they are. This is very unlikely and the general practice of fire wardens carefully and fully checking each floor during the evacuation process should cover this rare eventuality.

5 Visitors and customers

There is a difference in the way that an escape plan is provided where the person requiring the escape plan is a visitor to the building or is a customer using the service. The information required in Section 4 is still required for each group; however, it will not be possible to provide a bespoke plan for each person. Instead, a system of standard plans should be developed based on the matrix in Appendix 1.

Visitors should always be offered an escape plan, but staff should not be concerned if a person who has an apparent disability does not accept one. It is possible that the person is confident that they can make their own escape. This can apply to wheelchair users too (see Section 4.3: Wheelchair users). Members of staff should confirm with them that this is the case.

Generic plans should be provided in a discreet manner. This will encourage people who have conditions such as asthma, heart disease, epilepsy or emotional problems to ask for assistance, if they wish to do so. Their preferred escape method may be as individual as they are. However, it is likely to be met by one of the set standard PEEPs laid down for the building.

The service provider should adopt an approach that enables people to ask for a plan, when needed, without them feeling that it will affect the provision of the service to them. It should be understood that requesting a suitable evacuation plan would not result in restricted use of the building. All staff involved in the process of providing escape plans should be provided with a good standard of equality training to ensure that they do not inadvertently discriminate against disabled people.

In some public access buildings, such as museums, art galleries and shopping centres, there will be little or no control over the people who are present in the building. This can present a problem to the service provider. However, where a system of standard plans has been established, staff can be trained in the different escape options available. They can then be trained to offer an appropriate option to disabled people during an emergency and to lead them to appropriate points in the building.

Example 1
There are no step edge markings on the rear stair; however, the west stair has been provided with markings as part of building improvements. Both are available as escape routes. Staff should direct visually impaired people to the west stair.

Example 2

The building has a two-stage fire alarm system. The first stage is a coded message over the public address system to staff. At this point staff discreetly approach people who they consider may need assistance and ask them to leave prior to the confirmation of evacuation.

The matrix

The matrix in Appendix 1 includes most disability types and recommends options for their escape. When working in partnership with a disabled person to establish their escape plan, the matrix should be used as a guide to what options might be offered in the plan.

Assisted/facilitated escape options

This section explains each option shown in the matrix. In order to use the matrix, look at the escape option suggested for each disability type. The corresponding number in this section gives additional information on each type of escape. The two can be used together as part of the planning process for each person's PEEP.

The options can be used as a discussion tool in order to establish the options open to each person. They should be matched to each building, and one person's choice of escape may differ depending on the building. For instance, a visually impaired person may be able to find their way out of a building that has good orientation standards and is uncomplicated. However, in a complex building where there is poor signage and orientation they may need assistance.

This may mean that the person requires different plans for different buildings. Assumptions should not be made that the same plan suits all. Also, a disabled person should not be pushed into using the same method of escape in one building as they would use in a more accessible building.

1. Evacuation lifts

During a fire incident, once the Fire and Rescue Service is in attendance they will operate the lift override system to use the lifts themselves to access the fire. As a result, all lifts in the building to be used by the Fire and Rescue Service will return to the fire service access level and park. Once this happens, it will not be possible to call the lifts as they will be under the control of the Fire and Rescue Service.

Where suitable evacuation lifts are provided, disabled people should make their way to the lift point and use the communication system to contact the lift operator and make them aware of which floor they are waiting on. In addition, there will also be a refuge call point (adjacent to the evacuation lift) whereby the disabled person can contact the control room in order to tell control which refuge they are in.

Fire-fighting lifts may be used in the early stage of the evacuation process in agreement with the local Fire and Rescue Service.

In buildings where horizontal escape is used prior to exit in an ordinary lift, the instructions for horizontal escape should be followed first.

2. Meet assistance at a refuge

Some disabled people will require assisted escape. In these cases it will be necessary to have a pre-arranged meeting place. If the disabled person is likely to move around the building, a means of communication will be necessary between the escape volunteer and the disabled person. They can then arrange to meet at a particular refuge point during the escape.

People should never be left in a refuge point to wait for the Fire and Rescue Service. The refuge can be used as a safe resting place as well as a place to wait in a phased evacuation while the go-ahead for a full escape is established. A refuge may be equipped with a suitable means of communication.

Most refuges can accommodate only one wheelchair. This should not be a problem where there is more than one wheelchair user, provided that there is a suitable evacuation strategy in place. As one person progresses on their journey, the next person will take their place in the refuge. Fire compartmentation is also a form of refuge. The refuge may play a part in the disabled person's escape journey.

3. Meet assistance at a workstation

Some people will need to meet their assistant(s) at their own workstation. In this instance the allocated escape volunteer(s) should go straight to the disabled person's workstation at the beginning of the evacuation procedure. The assistant(s) could be someone who works alongside the disabled person (buddy system), therefore they can set off on their escape journey together.

4. Make own way down stairs slowly

Some people who use wheelchairs may be able to make their own way down the stairs if they have a little mobility. It may be necessary to ensure that there are suitable handrails and step edge markings present. The preferred solution is where the escape plan enables disabled people to leave the building by their own efforts. This reduces the chance of confusion and the chance of the plan breaking down. In these instances the person may rest along the way in refuges.

Disabled people who choose this independent method of escape are likely to move slowly down the stairs and it may be better for them to wait for the main flow of people to leave the building. Escape stairs that are incorporated in a fire-resistant shaft should be safe for up to 30 minutes. This greatly enhances the escape time, especially when fire alarm systems incorporate advanced fire detection measures. This reinforces the importance of building occupiers keeping self-closing fire doors shut and observing good housekeeping practices when occupying buildings.

Where this escape method is chosen, it is important that it becomes part of the PEEP and is recorded and monitored should there be a problem during the escape. The fire warden should report to the control room or the person in charge of the evacuation process that a disabled person is slowly making their way out of the building. This information must be passed on to the Fire and Rescue Service on their arrival at the incident.

5. Move down stairs on bottom after main flow

While some people will prefer to take responsibility for their own escape by walking down the stairs, others may prefer to make their own way out by shuffling down the stairs on their bottom. Again, it will be best for this group of people to wait until after the main flow of people has evacuated. Wherever possible, they should be monitored to ensure that there is no problem with their progress. The fire warden should then report to the control room or the person in charge of the evacuation process. This information must be passed on to the Fire and Rescue Service on their arrival at the incident.

6. Evacuation chairs

Where this is the preferred method of escape, the responsible person will provide an evacuation chair. It will be allocated to a particular person and either kept alongside their desk or in the most suitable refuge close to them.

In the case of a visitor who requests this method of escape, the person who is responsible for booking them into the building should contact the appropriate responsible person to arrange for one to be brought to the most suitable part of the building for the duration of their stay. It should then be returned to the central storage point.

In buildings with an uncontrolled and unknown population, it may be advisable to provide evacuation chairs at suitable points within the building. One on each staircase at each level may be an expensive option. Provision of evacuation chairs on the top floor of the building, with a communication system that allows them to be brought immediately to any refuge, may be an acceptable solution, depending on the fire safety measures in place and individual circumstances.

Provision of evacuation chairs should always be accompanied by a full system of escape for disabled people as they are only a part of the solution. Regular training of staff in the use of evacuation chairs is essential.

7–9. Carry-down

There are a number of types of carry-down techniques using two, three or four people. Where a disabled person wishes to be carried out either using their own wheelchair or by another method, a manual handling risk assessment should be carried out and a suitable team should be assembled and trained to take them out safely.

There are a number of pieces of equipment available to help with this evacuation technique. The appointed people require regular training to use any equipment safely. When carry-down is the preferred method, specialist moving and handling training should also be provided.

10. Move down stairs in own chair with support

Some wheelchair users are strong and skilled enough to tip their chair on its axis and travel down the stairs in this way. Others can do this with assistance. Where this method of escape is considered, expert training will be required and the technique should be practised regularly. Again, the escape should take place after the main flow of people leaves the building. It is only acceptable for short flights of stairs.

11. Cannot transfer readily

Some people will find it difficult or impossible to transfer from their chairs to an evacuation chair or other evacuation aid. These people may require a hoist to assist with this movement. The process can be quite difficult and suitable training is required. It may be appropriate, wherever possible, for a disabled person's workstation or a point of service used by disabled people to be located in a place where better evacuation plans can be made. In these cases, a risk assessment of the use of lifts within the building for evacuation purposes may find that this solution presents less of a risk.

Service providers could ensure that meeting or hotel rooms with easier evacuation routes are priority booked for disabled people who require a high level of assistance.

12. Move down stairs using handrails

Some people will be able to make their own escape but will require a handrail to support them to get out of the building. This will be to either the right or left of the stairs. Some will not be able to use the right and others the left. Once you have established that they require a handrail, check each staircase in the building proposed for their use to ensure that a suitable handrail is provided. Where one is not available then assistance may be required. Provision of handrails may be considered a reasonable adjustment.

13–14. Assistance from one to two people

Some people will require a buddy to assist them out of the building. Some will be happy to organise this themselves on a casual basis. If this is the case, a check should be made to ensure that the disabled person will always be in a group of their peers or regular staff who are able to provide this. If not, then it may be necessary to establish a formal procedure for times when they are likely to be alone. In these cases, it may be suitable for them to use the standard procedures set up in that building for visitors.

15. Orientation information

Where a person requires additional orientation information, it may be sufficient to give them a guided tour of the escape routes from the rooms they use. There are a number of disabilities where additional orientation information is required. Good orientation systems benefit all of them and could include colour coding, signage and defined routes (as explained in 16–18). People with cognitive impairments can benefit from a photographic record of the route.

16. Tactile maps of the building

Some people will need additional guidance information in the form of tactile maps. These can be obtained through a number of organisations that provide accessible information services. You should ensure that you have this information in advance.

17. Colour contrasting on stairways

To assist their orientation needs, some people will require an orientation strategy to assist way-finding within the building. This can be achieved through colour coding or contrasting the escape routes. An alternative for smaller organisations might be to provide laminated paper signs with red triangles and yellow squares printed on them; these are used to identify the escape routes and supplement the regulated escape signs.

18. Step edge markings

Some people will be more confident about making their own way out of the building if there is sufficient contrast on the nosings on the stairs. If a person requests this option, the stairs should be checked to see if the step edges are highlighted. If they are not, the person may require a buddy to help them out of the building. It may also be appropriate to allocate a working area close to where there are suitable step edge markings. It would be advisable to provide contrasting nosing on all stairs in order to reduce the need for assistance.

19. Need to be shown the escape routes

Some people will only need the escape routes pointing out to them and this will be sufficient.

20. Assistance for the person and their dog

Where a person uses a guide dog, they may prefer the dog to assist them out of the building. The escape routes should be pointed out to them. Others will prefer to take the responsibility away from the dog for means of escape and request a human assistant. In these cases, a buddy should be allocated to the person. It may also be necessary to provide a person to look after the dog. Again, this may be provided in an informal or formal manner.

21. Need doors to be opened

Some people may have difficulty negotiating self-closing fire-resisting doors. It should be ensured, therefore, that all such doors and their self-closing devices (including those that are normally held open by electromagnets linked to the fire alarm system) comply with the recommendations of the appropriate British Standard regarding opening and closing forces.

However, some people may still require assistance to open the doors, for example those with upper limb impairments. Again, this can be a formal or informal arrangement. Where a person may be alone in a building that has doors that may be difficult for them to open, it may be necessary to provide a more formal level of assistance.

Managers should ensure that a fire door self-closing device is not set at too strong a pressure and they may need to adjust it, but it must also be remembered that such doors are designed to hold back smoke and fire to protect all the people in a building and facilitate their escape.

22. Large print information

Some people will need fire evacuation information provided in large print. This can be obtained through a number of organisations that provide accessible information services. Alternatively, it is possible to produce large print information in-house. Ask the person what size of print is suitable for them. You should ensure that you have this information in advance.

23. Identification of escape routes by reception or security

Visitors to the building may need reception or security staff to show them the escape routes when they arrive at the building. This task should be allocated to the most suitable person for each building or department.

24. Flashing beacons

Hearing impaired or deaf people need to be made aware that an evacuation is taking place. Where they are likely to be alone in the building, they may need to be provided with a flashing beacon or other similar device. If this type of system is required, check with the appropriate person to see if there is one available within the building. Where there is not, then a suitable buddy system will be required. Flashing beacons may not be appropriate in all buildings, for instance where other lighting conflicts with the beacons.

25. Buddy system

A buddy system may be the most suitable method for alerting a hearing impaired or deaf person to the operation of the fire alarm. This should not be done on an informal basis in case everyone assumes that someone else has given the warning.

26. Vibrating pagers

Vibrating pagers can alert hearing impaired and deaf people that there is an emergency and they need to leave the building. They can also be used to communicate with other people who are part of the assisted escape system. The pagers can be used to inform people that there is a need to escape and also to tell them which direction they should travel in.

27. Alternative alarm systems

There are other methods of contacting disabled people; these can be either through the telephone system or through the intranet. It is recommended that where a person cannot use the existing system or needs support to use it, all other communication options are explored.

28. Additional checks by fire wardens

In order to provide back-up wherever there is an assisted escape system in place, it is also necessary for the fire wardens or fire marshals in the building to be aware of who is present and what escape plans are in place. They should then be trained to provide suitable assistance where necessary.

29. Horizontal evacuation

In some buildings, it is possible to evacuate people horizontally through the building into another fire compartment and away from the emergency situation. When the alarm goes off, people who cannot use stairs are directed to move along the floor level they are on to another fire compartment.

Information about where to go is required in order for this system to work. Where horizontal evacuation is not immediately available on the affected floor, it may be available on a lower floor. This may be more acceptable than travelling all the way to the ground floor. The opportunity to do this should be identified as part of the building fire safety risk assessment and then offered during the interview.

30. Taped information

Where a person cannot read the fire drill instructions, they may benefit from their provision in tape format. This should be produced in Plain English and in other languages where appropriate.

Appendix 1 – The matrix

Option	Type of escape	Electric Wheel-chair user	Wheel-chair user	Mobility impaired person	Asthma & other breathing/ health issues	Visually impaired person	Hearing impaired person	Dyslexic/ orientation disorders	Learning difficulty/ autism	Mental Health problems	Dexterity problems
1	Use of lift	●	●	●	●						●
2	Meet assistance at refuge		●	●		●				●	
3	Meet assistance at work-station	●	●	●	●	●	●	●	●	●	●
4	Make own way down stairs slowly	●	●	●	●	●					
5	Move down stairs on bottom after main flow	●	●	●	●				●		
6	Use evacuation chair or similar	●	●	●	●						
7	Carry-down 2 people	●	●	●	●						
8	Carry-down 3 people	●	●	●	●						
9	Carry-down 4 people	●	●	●	●						
10	Travel down in own chair with support		●								
11	Cannot transfer readily	●	●								
12	Can get down stairs using handrails	●	●	●	●	●			●		
13	Needs assistance to walk down stairs 1 person	●	●	●	●	●				●	●
14	Needs assistance to walk down stairs 2 people	●	●	●	●	●			●		
15	Need orientation information					●	●	●	●	●	

Option	Type of escape	Electric Wheelchair user	Wheelchair user	Mobility impaired person	Asthma & other breathing/ health issues	Visually impaired person	Hearing impaired person	Dyslexic/ orientation disorders	Learning difficulty/ autism	Mental Health problems	Dexterity problems
16	Needs tactile map of building					●					
17	Need colour contrasting on stairways					●		●	●	●	
18	Needs step edge markings			●		●	●	●		●	
19	Needs showing escape routes				●	●	●	●	●		
20	Needs assistance for person and dog				●	●					
21	Needs doors opening										●
22	Large print information					●		●	●		
23	Identification of escape route by reception/ security					●	●	●	●	●	
24	Provisions of flashing beacons						●				
25	Buddy system					●	●	●			
26	Provision of vibrating pagers	●	●	●			●				
27	Provision of alternative alarm	●	●	●		●	●				
28	Additional checks by fire wardens						●	●	●	●	
29	Horizontal evacuation	●	●								
30	Need for taped information					●		●	●	●	

Appendix 2 – Pro-forma letter

Dear

Personal Emergency Evacuation Plans (questionnaire)

We are currently reviewing and improving our emergency evacuation procedures and we want to ensure that all of our staff are able to leave the building safely in the event of a fire or other emergency. We understand that many disabled people will be able to leave the building unaided; however, some may require assistance. Therefore, we are writing to you to ask you whether you would like us to draw up a Personal Emergency Evacuation Plan (PEEP) with you in order to ensure that you can leave the building safely in the event of an emergency.

The plan will explain what options you wish to take in the event of a fire evacuation. The plan will also state who is designated to assist you in your escape should you require this. The human resources department or other manager, in full consultation with you, will draw up your PEEP. These people will have been trained on disability equality issues and will work with you to find the best solution.

We are including a questionnaire for you to fill in to help you assess your own need for a plan. Please return the questionnaire as soon as possible/by .... If you do require a plan, we will arrange a meeting with you to discuss it. If necessary, we will appoint people to help you. You will receive a copy of your plan, which will also be given to those people who are part of your escape plan. The fire incident controller (or other) will also receive a copy and will pass it on to the Fire Service if necessary. If you do not request a plan, we will accept that you are able to make your own way out unaided.

This does not affect your right to employment. As your employer we have a duty to provide you with a suitable escape plan regardless of your disability. We will not expect you to make any extraordinary effort to escape at any other time.

If you have a temporary condition that may impede your evacuation, such as pregnancy, please inform us if you feel you need assistance. If your disability does not normally affect your work but might be a problem in an escape situation, please inform us so that we can arrange suitable assistance. This will not affect your right to employment.

Thank you for taking the time to fill in the questionnaire, which will enable us to bring about any necessary changes.

Yours sincerely

Appendix 3 – New starter evacuation questionnaire

Have you read and understood the evacuation procedure for the building in which you work?

Yes   No

Do you require the procedure in large print or in another alternative format?

Yes   No

If yes, please state which: _____

Do you have any special evacuation requirements?

Yes   No

We operate an evacuation system that includes Personal Emergency Evacuation Plans (PEEPs) for disabled staff. If you have answered yes to the above question, you will shortly receive a questionnaire.

Please fill it in as quickly as possible and return it to_____

If you have any questions, please speak to_____

Thank you

Appendix 4 – Personnel record sheet

Name	Department	Evacuation plan
Mike Smith	Engineering	Plan 15
Jake Long	Maths	Plan 5

Appendix 5 – PEEP option 1

Part 1 _____

Name: _____

Location: _____

Alternative working positions (if appropriate): _____

Location: _____

Indicate the number of separate plans that have been provided for each building and room visited.

Building name   Room numbers

_____

_____

_____

_____

Part 2: Awareness of procedure

I have received the evacuation procedure in the following format:

- Braille
- Electronic format
- Tape
- Large print
- It has been explained in BSL
- I have been shown the evacuation routes
- I have my own authorised plan

Alarm system

- I am informed of the emergency by:
- The existing alarm system
- Pager device
- Visual alarm system
- Members of my work team
  (*each of these people require a copy of this sheet*)
- The fire wardens on my floor (*the fire wardens require a copy of this sheet*)

Names:_____

Part 3: Getting out

I require ___ people to assist me.

Names:_____

Back-up:_____

Each of these people require a copy of this sheet.

The following is a record of my escape plan:

_____

_____

_____

_____

Each of these people require a copy of this sheet.

My specialist equipment to assist my escape is:

_____

My practice diary is:

| Year 1 | | | | | | | | | | | | Year 2 |
Jan	Feb	Mar	Apr	May	Jun	Jul	Aug	Sep	Oct	Nov	Dec	Jan
MOE training	Carry-down training	Mock-up				MOE training	Carry-down training	Mock-up				MOE training

Date:_____

Example of evacuation procedure

This is a step-by-step account of what will happen during the escape.

John and Gale *will meet me at my desk.*

Reserve volunteers are ***Maria and Mike***

They will help me by taking hold of one arm each side.
We will walk to the nearest escape route and wait in the space at the head of the stairs for other people to escape.
When it is safe to do so, we will move slowly down the stairs.
The fire warden will advise the Fire and Rescue Service which route we took.

Appendix 6 – PEEP option 2 (simple record sheet more relevant for standard plans)

Option 15

Requirement

My sight is limited and orientation is difficult where there is no formal guidance.

Escape procedure

The person you are visiting will take you to the refuge, which is within the evacuation stairway at each level of the building.

Please ring for assistance from the call point situated within the refuge. A member of our fire evacuation team will meet you there and guide you out of the building.

A more suitable variation on this is where all staff are trained to assist visually impaired people out of the building.

Specialist equipment to assist the escape is:

_____

Fire warden checks

Communication point

Option 8 – Carry-down by two staff

Requirement

I can walk on the flat but cannot manage stairs at all. I would need to be carried down the stairs.

Escape procedure

Please make your way to the refuge, which is within the evacuation stairway at each level of the building. Please ring for assistance from the call point situated within the refuge.

Our staff are trained to carry-down with the use of an evacuation chair and two staff.

A team will meet you in the refuge. You will need to sit on the chair, which has armrests to help support you. The two staff members will then carry you down.

Specialist equipment to assist the escape is:

_____

Evacuation chair

Appendix 7 – Reception sign

Option 1 – Standard PEEPs in place

> **Option 1 – Standard PEEPs in place**
>
> We operate a system of assisted escape for disabled visitors.
> Please tell our receptionist your requirements.
>
> We will provide you with a suitable escape plan.

Option 2 – Disabled people's evacuation strategy in place

> **Option 2 – Disabled people's evacuation strategy in place**
>
> We operate a system of assisted escape for disabled visitors.
> Please tell our receptionist your requirements.
>
> We will explain our escape procedures to you.

Glossary

Term	Definition
British Sign Language (BSL)	Form of sign language developed in the United Kingdom for the use of the deaf. Indigenous language.
BS 8300	British Standard 8300: 2001 on Design of buildings and their approaches to meet the needs of disabled people. Code of practice.
Competent person	A person with enough training and experience or knowledge and other qualities to enable them to properly assist in undertaking the preventive and protective measures.
Disability Discrimination Act 1995	Legislation passed in 1995 to address discrimination against disabled people
Disability Equality Duty	The Disability Equality Duty came into force on 4 December 2006. This legal duty requires all public bodies to actively look at ways of ensuring that disabled people are treated equally.
Matrix	Table or grid (in Appendix 1, used to assist in ascertaining appropriate means of escape)
Personal Emergency Evacuation Plan (PEEP)	Individual plan for means of escape from fire.
Plain English	Writing that the intended audience can read, understand and act upon the first time they read it.
Regulatory Reform Order (Fire Safety) 2005 (RRO)	Legislation on fire safety for non-domestic premises.
Responsible person	The person ultimately responsible for fire safety as defined in the Regulatory Reform Order (Fire Safety) 2005.

Index